SCULPTURE BEAUTY'S

人物模型之美

辻村聰志　女性人物模型作品集

SATOSHI TSUJIMURA GIRLS FIGURE WORKS

辻村聰志

PROLOGUE

欣賞人物模型作品時，大家的視線會先落在何處？我個人的順序是先看人物的臉，接著看整個模型。這本作品集幾乎收錄了我在長谷川製作的所有擬真人物模型。雖然不知道各位翻閱此書時視線會落在何處，但是我希望大家在翻閱時不會有「這裡待加強」的感覺。這是因為在提交原型時，這不但是通過我心中標準的作品，也是取得長谷川認可的成品，然而經過一段時間，依舊深感自己的能力尚未成熟，覺得「如果如此這般就更好了」。這表示我製作的人物模型還稱不上完美。希望大家看了這本書，可以依照自身的喜好塗裝和改造，彌補這些遺憾。

辻村聰志

Each person has their way of looking at figures. For me, the very first thing I see is their face and then their entire body. This book covers most of the realistic style figures I created for Hasegawa. I'm not sure what and where you're looking at when reading this book, but I sincerely hope you are satisfied with the quality of my figures. Of course, I sculpted every figure listed in this book to meet the quality requirement of Hasegawa. I can safely say that I put the maximum effort into creating each figure, but as time goes by, I realize how inexperienced I was back then. Thoughts like "I should have done it this way" pop up quite frequently, and in that sense, my figure creations are far from perfect. I hope you will be able to make up for that imperfection by better understanding my figures through this book and by painting and modifying each figure to your liking.　Satoshi Tsujimura

1971年出生於東京，居住於三重縣。自幼稚園起就喜歡塑膠模型，在小學高年級時，就喜歡使用琺瑯塗料塗裝1/700比例的船艦模型和城廓模型。在國中三年級即將畢業時將興趣轉向繪畫，短暫遠離喜愛的模型製作，直到30歲左右因為舊時好友的關係再度牽起與模型的緣分，兩人原本在談論學生時代製作模型的回憶，後來變成個人委託人物模型的製作，就這樣又回到了模型的世界。之後一邊過著一般的上班生活，一邊承接人物造型的工作，直到過了40歲獨立成為自由接案的原型師，開始製作MODELKASTEN 1/35比例女高中生樹脂人物模型的原型。2015年起負責製作長谷川商品，其中1/24、1/12和1/35比例等樹脂鑄造的人物模型和塑膠模型獲得許多消費者的支持，成了長谷川的明星商品之一。

辻村聰志
SATOSHI TSUJIMURA

Tsujimura was born in 1971 in Tokyo, Japan, and currently resides in Mie prefecture. He has enjoyed building plastic models since kindergarten. His preferred genre of models were 1/700 scale ships and Japanese castle models. At the end of his third year of junior high school, his focus gradually shifted towards painting artworks, but he eventually returned to the hobby in his early 30s. Since then, he has been sculpting figures in his spare time. After turning 40, he became a full-time freelance figure sculptor and worked on projects such as 1/35 scale high school girls resin figures and Hasegawa's 1/24, 1/12, and 1/35 figure series.

INDEX

PHENOMENON

夢幻合作始於長谷川的「認真看待」

The start of a miraculous collaboration between Tsujimura and Hasegawa

撰文／寒河江雅樹　Text：Masaki Sagae

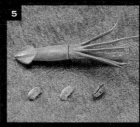

1 贈品人物模型「莉狄亞·利特維亞克」，為辻村聰志與長谷川初次合作的「擊墜王－蒼空7人－（第2次世界大戰王牌飛行員和7架戰機套組）」配件。

2 3 繼飛行員人物模型之後，原型製作選用的題材為操作「雙臂挖掘機ASTACO NEO」或「鋼輪全液壓振動壓路機」（兩者都是長谷川1/35比例的模型）建築機械的1/35比例女性操作員。

4 「建設作業員套組B」（長谷川1/35比例的模型）。這也是辻村聰志製作的原型。技巧純熟，能製作出男女差異的作業員。

5 辻村聰志還負責製作了「深海6500號」（長谷川1/72比例的模型）限定版套件中的大王烏賊和大王具足蟲。除了擬真人物模型之外，還會製作這類題材的原型。

1. One of Tsujimura's first sculptures for Hasegawa.
2.3. These operators were included in Hasegawa's construction equipment series released since 2015.
4. Figures from 1/35 construction worker set.
5. These deep-sea creatures are also the works of Tsujimura.

　　長谷川為比例模型商品廠商，以飛機為主推出許多商品系列，並且在2014年推出一項套組銷售企劃，就是立體製作第2次世界大戰中，7名來自不同國家的王牌戰鬥機飛行員和戰機。當中部分的人物模型原型就是由辻村聰志負責製作。即便是1/48比例的極小模型，但是造型精緻，可充分感受到女性的柔美樣貌和表情，成品的質感之高受到長谷川的注意，而在2015年持續委託辻村聰志製作1/35比例建築機械套件中，操作車輛的女性操作員人物模型。建築機械和女性操作員的特殊搭配，加上從任何角度觀看都非常可愛的人物模型造型，受到市場極大的好評。之後在2017年為了推出1/24比例汽車模型搭配女性人物模型的組合商品，長谷川又請辻村聰志製作商品原型，大家對此應該都還記憶猶新吧！

　　過往的擬真女性人物模型和辻村聰志製作的人物模型有很大的區別。其不同之處在於辻村聰志製作的人物模型即便尚未塗裝，完成度之高甚至讓人覺得「已經夠可愛了」。過去的擬真人物模型，或許是造型的關係，大多讓人覺得有種無法言喻的恐怖感，尤其是未塗裝的人物模型更為明顯。過去模型師內心都暗自想像「希望有個可愛的女孩站在自己做的機械模型旁邊」，對他們來說，由於自己不擅長塗裝，有個不需塗裝只擺放一旁就很可愛的人物模型真是太棒了。而長谷川依照市場反饋也順勢給予正向的回應，推出了可以擺

放在汽車模型旁邊的人物模型，模型不是樹脂製，而是射出成型的塑膠模型。這次推出的塑膠模型一口氣提升了長谷川擬真人物模型的知名度，進而開始發展1/24比例甚至1/12比例等多種類型的商品系列。就這樣現在長谷川以幾乎每月都會推出的超高頻率，銷售由辻村聰志原型製成的商品。這個現狀證明了辻村聰志的造型實力之高完全切中了消費者的需求，更重要的是這代表長谷川本身也深知辻村聰志人物模型的魅力，並且判斷可將其納入一項商品類型。而最令人驚嘆的莫過於辻村聰志的製作才能，他不斷以不可思議的高速製作原型，以因應長谷川或是說更前端消費者對於系列商品的需求。而且他還能精準掌握主題設定的角色背景、人種和風俗，完美刻畫出每個人物模型並且賦予個性，擬真還原的能力非常高，很值得受到大家的讚許。

　　有很多人透過長谷川商品認識辻村聰志這位原型師，並且感受到其造型創作的魅力，當然也不少人是因為這本書才知道辻村聰志的名號。我想要再次表達的是長谷川屬於「寫實」路線的比例模型製造廠商，但卻認真投入人物模型的商品製作，其實這本身已是一大新聞並且令人感到不可思議。本書想強調長谷川的認真看待和擬真人物原型師辻村聰志之間的合作，創造了「夢幻般的人物模型」，希望各位讀者可從書中品味其完美搭配的魅力。

Traditionally, Hasegawa's focus has always been around scale models, especially aviation models. So it was no surprise that the first collaboratory product with Tsujimura was series of seven ace pilot figures, each added as a bonus to limited-edition World War II fighter kits. These were followed up by female operator figures attached to 1/35 scale construction equipment kits, which received a great response from the market. There is a rather significant difference between contemporary female figures of the past and Tsujimura's figures. His figures, without even painting, already look cute with their delicately modeled postures which capture feminine beauty perfectly, as well as their facial expressions. The demand for realistic yet attractive female figures that can be displayed next to your car model was high, and Tsujimura/Hasegawa collaboratory kit's popularity quickly increased in a short period. For many readers, this might be the first time you ever heard of a sculptor named Tsujimura. You might also be surprised to know that Hasegawa, a "traditional" scale model manufacturer, has put a lot of effort into producing female injection/resin figures. This book aims to introduce readers to the mesmerizing and fascinating world of Tsujimura through his realistic female figures, which were made possible by collaborating with Hasegawa.

1/12比例擬真人物模型系列

1/12比例的尺寸比1/24比例的人物模型大，在原型製作的作業量、材料用量上也明顯較多，老實說是非常需要體力的工作，甚至在補土塑形就會讓手很痛。不過因為比例較大，所以可以添加相應的細節設計，的確也使創作充滿樂趣。最近數位造型興起，非常方便於細節和尺寸的調整，但是我依舊習慣過去的人工造型作業。因此細節各處仍然以手工製作，所以相當耗費心神。當然作業時會使用頭戴式放大鏡，但是即便是這個尺寸仍然覺得有其限制。不過另一方面，依照所想完成這個比例尺寸的作業時，也能獲得極大的喜悅。

Compared to 1/24 scale, one clear thing is that 1/12 scale figures are a lot bigger. 1/12 scale translates to more work and more materials required for the sculpture, to the point that my hands get sore after kneading a massive blob of putty. However, the larger the scale, the more fun I can have by incorporating more details and more delicate lines. In terms of adding minuscule details, digital sculpting has its upper hand. However, I am an old-fashioned sculptor who prefers to use his own hands. Naturally, I use a magnifier when working, but I sometimes feel limited even at a larger scale, such as 1/12. At the same time, the pleasure I get when things go just as planned is incredible.

雖然主題是騎士，但是穿著較為輕便，造型給人夏
季穿搭的感覺。因為是初次製作1/12比例的系列，
在原型製作上還不熟悉尺寸的拿捏。為了瞭解眼睛
大小和臉部五官比例，造型製作時還特地拿尺仔細
丈量尺寸。豐富的姿勢變化，隱約呈現出身體的性
感曲線。

12 REAL FIGURE COLLECTION No.01
"GIRL'S RIDER"

1/12比例擬真人物模型系列No.01
「女騎士」（2020年）

雖然主題是騎士，但是穿著較為輕便，造型給人夏
季穿搭的感覺。因為是初次製作1/12比例的系列，
在原型製作上還不熟悉尺寸的拿捏。為了瞭解眼睛
大小和臉部五官比例，造型製作時還特地拿尺仔細
丈量尺寸。豐富的姿勢變化，隱約呈現出身體的性
感曲線。

This was the first figure I made for the "12 Real Figure
Collection." I initially had trouble getting a good
sense of the size of the figure, so I ended up using a
ruler to make sure I have the correct measurements for
the majority of the sculpting process.

12 REAL FIGURE COLLECTION No.02
"BLOND GIRL"

1/12比例擬真人物模型系列No.02
「金髮女郎」（2020年）

系列第 2 彈，我記得當初突然收到泳裝造型的委託需求時，覺得這個法宗響挺大膽冒險。另外我響不熟悉比例掌握，作業時將女騎士橫放當成參考。因為露出較多肌膚，所以特別注意零件分割線的設定位置，以及分割後再次接合處的造型不能顯得粗糙。

I remember feeling quite adventurous when Hasegawa asked me to create a swimming suit motif for the second product in the series. I had to be extra careful where and how to divide each component to hide part-joints cleverly.

12 REAL FIGURE COLLECTION No.03
"PADDOCK GIRL"

1/12比例擬真人物模型系列No.03
「賽車女郎」（2020年）

人物模型的姿勢如實表現出賽車女郎在鏡頭下刻意
露齒微笑的誇張表情。造型製作時特別留意要呈現
漂亮的身形，並且讓腳顯得修長。服裝的質感是參
考了各種資料所挑選設定。委託指示中表示要在縫
線部分塗上不同的顏色，所以還特別用刻紋表現出
來。

I hope you can sense her confidence through her
smiles. Her legs are sculpted a tad bit longer to make
her look stylish, just like a real race queen. The
stitching detail of the costume was carved out and
painted in a different color.

12 REAL FIGURE COLLECTION No.04
"BLOND GIRL VOL.2"

1/12比例擬真人物模型系列No.04
「金髮女郎Vol.2」（2021年）

泳裝設計特別，以參考資料為基礎再和長谷川討論
後設計而成。兩側半透明的材質部分使用珍珠塗料
表現出透明的獨特質感。雖然有考慮不要將頭髮太
往左右散開，不過自覺成品呈現的比例還頗為協
調。

Her swimsuit has a very distinctive design, which was
finalized after several discussions with the folks at
Hasegawa. I used pearl paint for the translucent
material on the sides of her swimming suit to add
more detail to the overall figure.

12 REAL FIGURE COLLECTION No.05
"GRAVURE GIRL"
1/12比例擬真人物模型系列No.05
「封面女郎」（2021年）

製作的人物風格設定為在海邊遊玩，充滿活力的現代偶像。主題為日本人，所以整個體型較為嬌小，並且留意在臀形等立體部分呈現出不同於過往外國人的作品表現。背影也調整成一眼就可知是日本人的身形比例。雖然為短髮，但是稍微添加一些分量，呈現微風將頭髮吹起的樣子。

I tried to make her look like a modern-day idol playing cheerfully on the beach. This figure's height is shorter since she is Japanese, and I made sure to sculpt her body features, such as her hips, differently compared to previous figures.

12 REAL FIGURE COLLECTION No.06
"BLOND GIRL VOL.3"

1/12比例擬真人物模型系列No.06
「金髮女郎Vol.3」（2021年）

不同於過去的金髮封面女郎，將人物設定在戶外現場展會的氛圍。身穿夏季服裝，身材比例較為修長，散發時尚氣質。我自己在為金髮塗裝時，不會混色而是直接塗上金色，並且會再以黑色＋透明紅的混色入墨線，讓頭髮更漂亮。

This figure gives a slightly different vibe compared to my other works, as if she is participating in an outdoor live music festival. Her outfit is summery, and her proportions are slim and stylish. Gold paint was used for her blonde hair.

12 REAL FIGURE COLLECTION No.07
"GRAVURE GIRL VOL.2"

1/12比例擬真人物模型系列No.07
「封面女郎Vol.2」（2021年）

這個封面女郎雖然同樣是日本人，但是身形較為修長而且為動態姿勢。上半身往前傾，所以製作原型期間要不斷確認人物是否可以自行站立。我覺得這個姿勢除了適合搭配相同比例的機械模型之外，也可以透過小物件的搭配組合，就可以呈現充滿故事的畫面。

Since this figure is leaning slightly forward, I tried to check many times during the sculpting process to make sure she could stand on her own. Adding small accessories such as beachball would make this figure more lively.

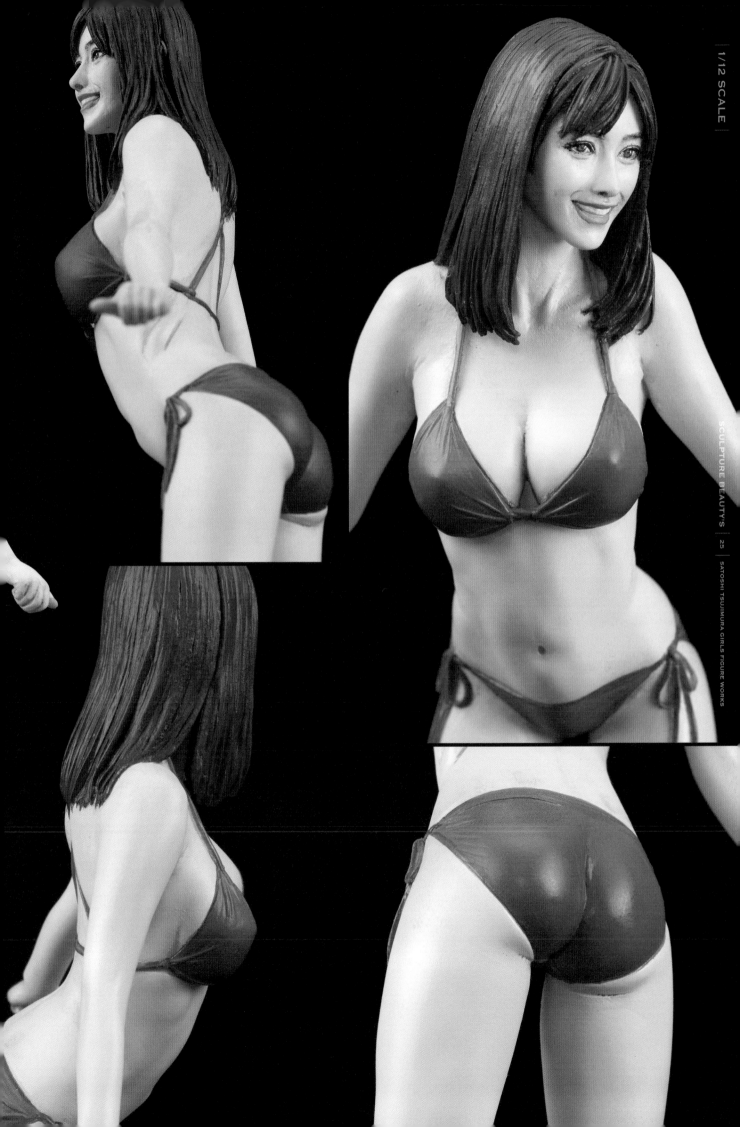

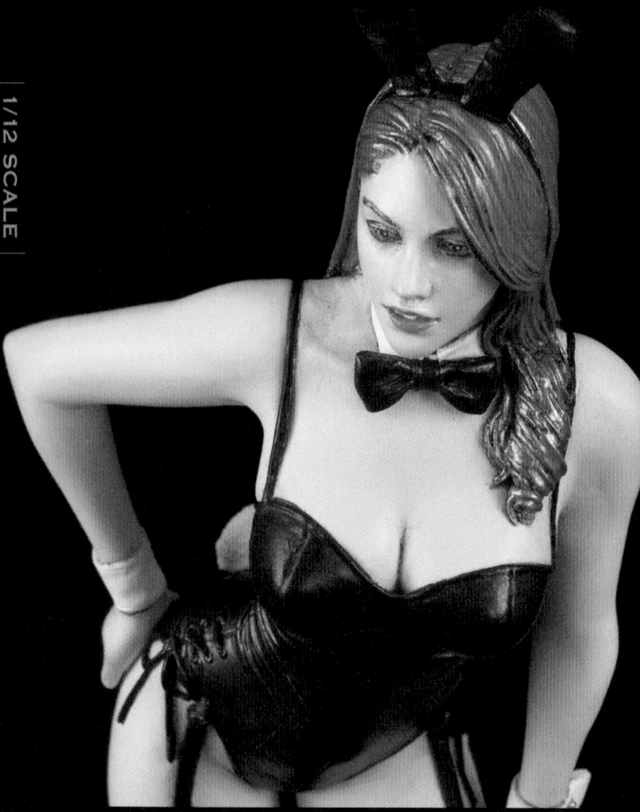

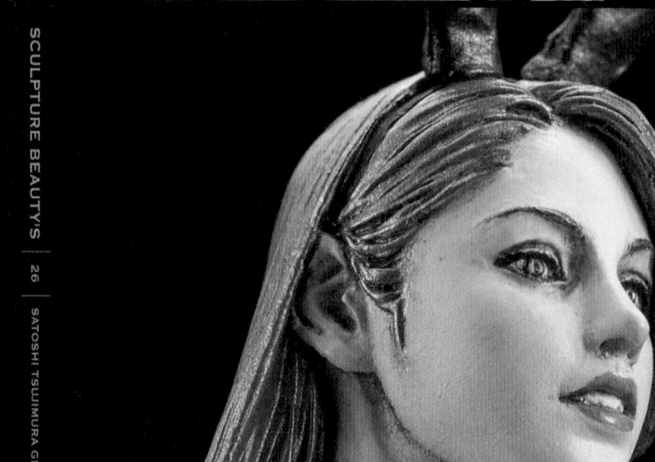

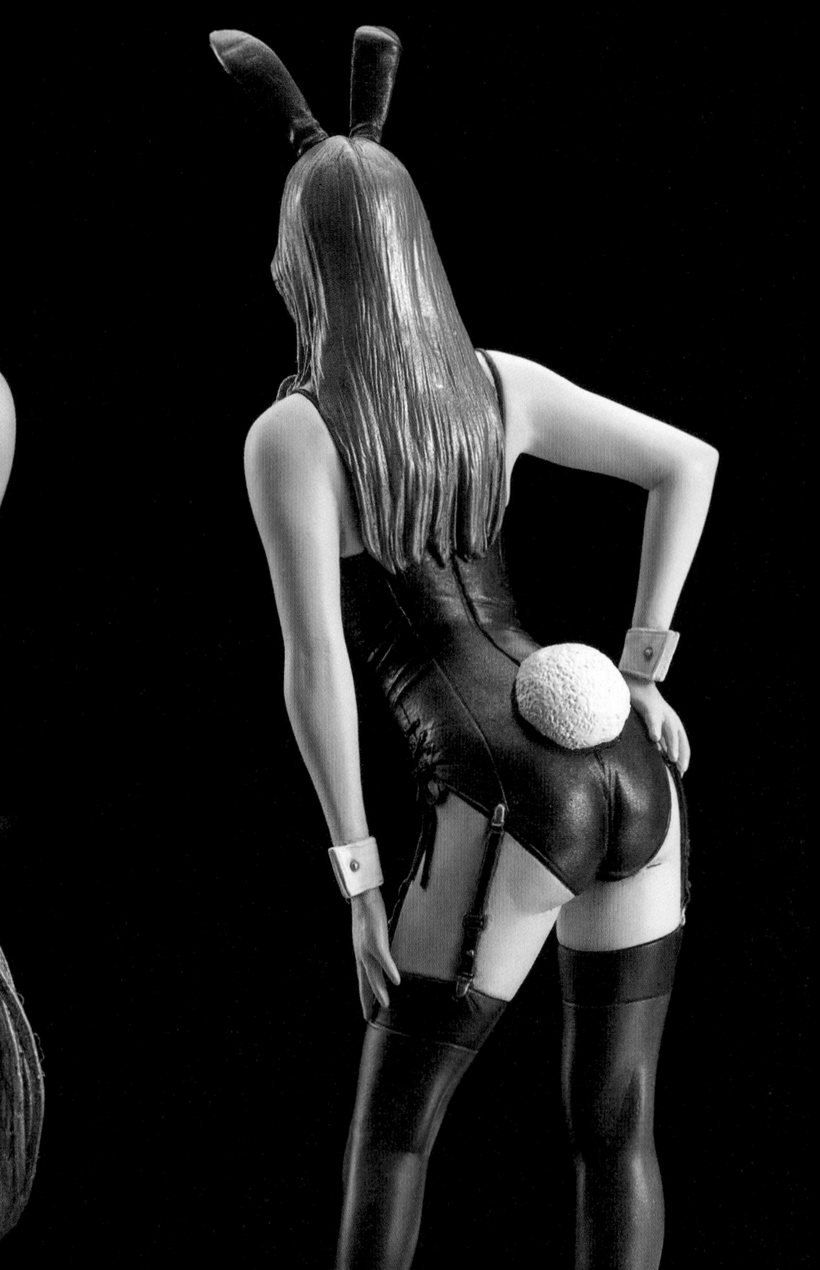

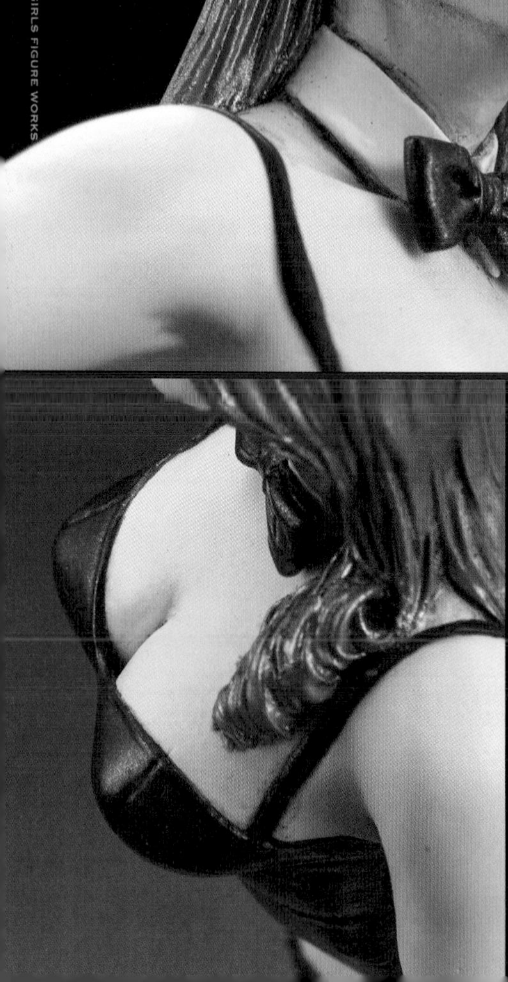

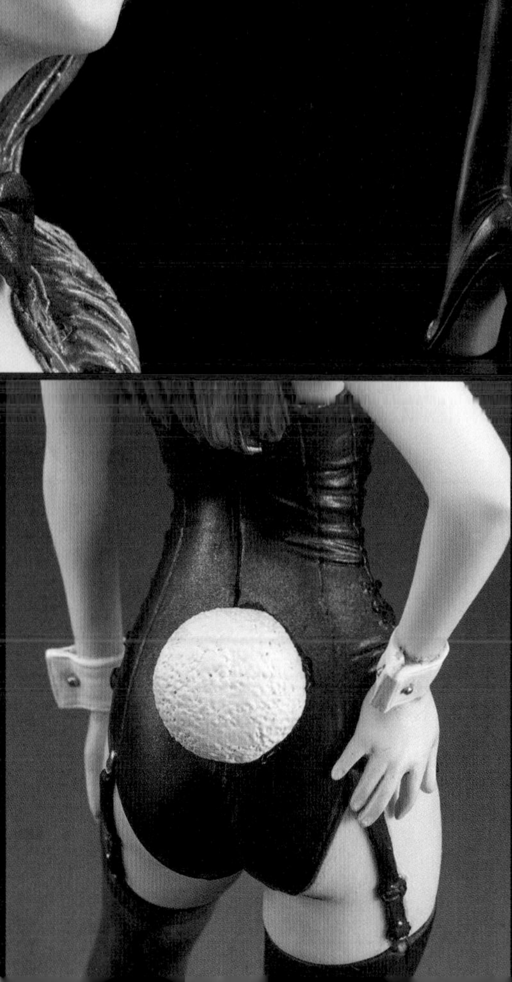

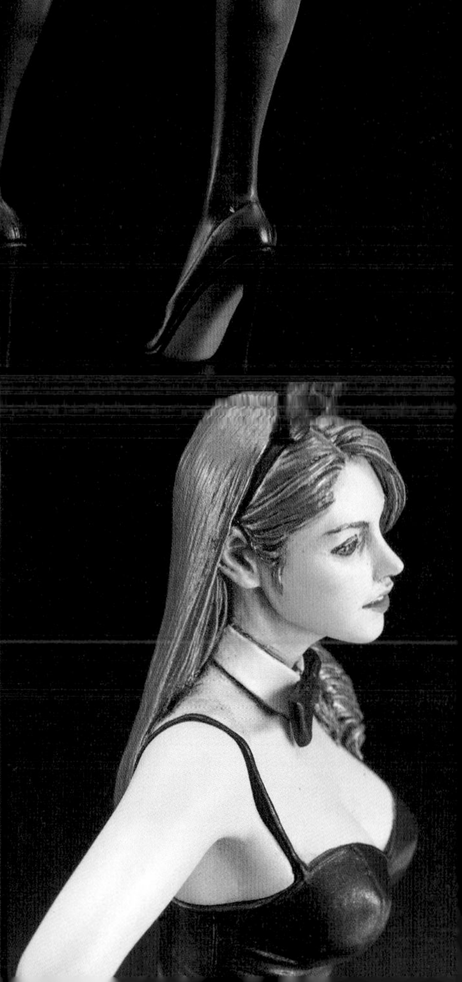

12 REAL FIGURE COLLECTION No.08
"BLOND GIRL Vol.4"
1/12比例擬真人物模型系列No.08
「金髮女郎Vol.4」（2021年）

雖然我沒有去過拉斯維加斯，但是人物造型就像拉斯維加斯賭場會出現的花花女郎。姿勢和表情有點撩人，就是當客人拍照時會立刻擺出與肩同寬的步伐和手插腰間的樣子。服裝質感不同於泳裝，所以用珍珠塗料將整件衣服鍍膜。

She looks like a play-girl right out of the Las Vegas casino, doesn't she? Of course, I have never been to a real casino, but you get the idea. The bunny suit is coated with pearlescent color to give its iconic texture and shine.

12 REAL FIGURE COLLECTION No.09
"HOSTESS"

1/12比例擬真人物模型系列No.09
「酒店女公關」（2021年）

刻意設計成第一眼就讓人感到華麗的樣子。為了精
緻做出有規則的首飾，讓我耗費心力。頭髮高高盤
起的細節也是造型亮點。雖然沒有實際走訪店家，
但考慮妝容不要太濃，眼線也只有稍微添加，讓人
物更端莊一些。

This is one of my latest works. Sculpting tiny details
such as her neckless was quite tricky. Again, I've
never been to a cabaret club, but I assumed their
make-ups wont be that heavy. So, I only emphasized
her eye-line and kept it simple.

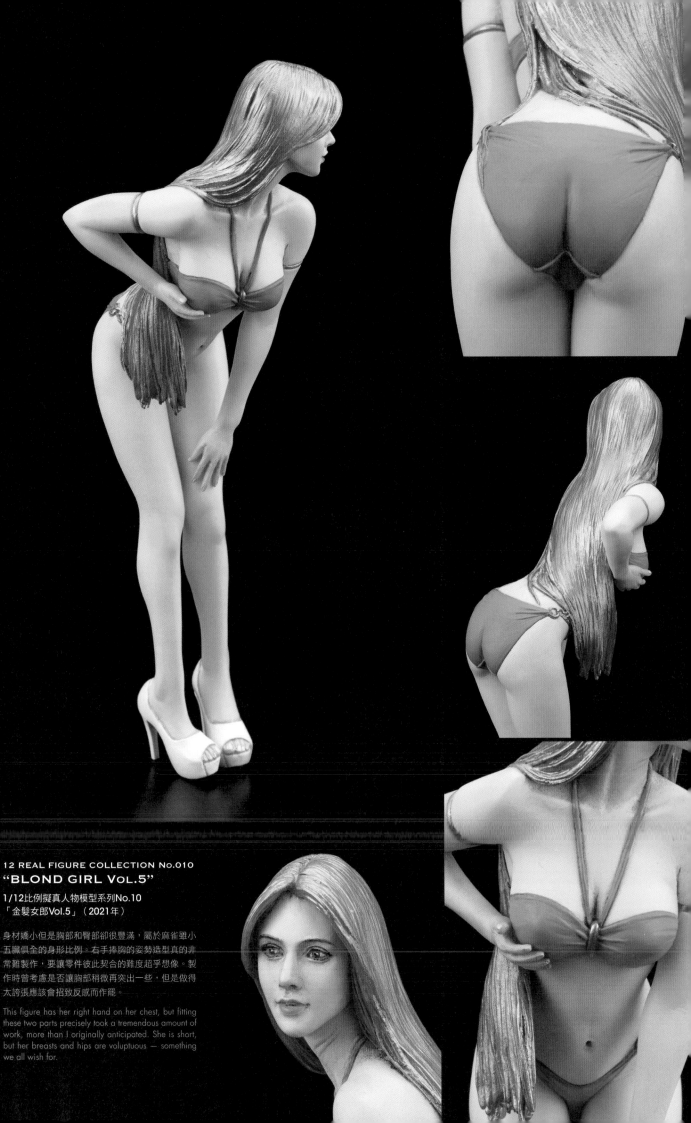

12 REAL FIGURE COLLECTION No.010
"BLOND GIRL VoL.5"

1/12比例擬真人物模型系列No.10
「金髮女郎Vol.5」（2021年）

身材嬌小但是胸部和臀部卻很豐滿，屬於麻雀雖小五臟俱全的身形比例。右手捧胸的姿勢造型真的非常難製作，要讓零件彼此契合的難度超乎想像。製作時曾考慮是否讓胸部稍微再突出一些，但是做得太誇張應該會招致反感而作罷。

This figure has her right hand on her chest, but fitting these two parts precisely took a tremendous amount of work, more than I originally anticipated. She is short, but her breasts and hips are voluptuous — something we all wish for.

ROOTS PART 1

成為造形師的基礎 「水彩肖像畫」和「中村主水」
The very beginning of Tsujimura's sculptural journey

撰文／寒河江雅樹　Text : Masaki Sagae

▲國立歷史民俗博物館收藏的江戶橋廣小路模型（2007年製作）。展覽初期的250個人物模型是由辻村聰志負責，而且他在短短 3 個月的時間內即製作完成。

A miniature model of an Edo-period city exhibited in the National Museum of Japanese History. Roughly 250 figures in this miniature were all built by Tsujimura.

　　辻村聰志和其他擬真人物模型原型師最大的區別在於，他正式踏入人物模型造形領域的曲折經過。
　　他在國中一年級春天時看了「風之谷」後，讓他有了「想描繪寫實繪畫」的念頭，自此超過15年以每天一幅畫的速度持續描繪水彩畫，以滿足內心對於繪畫的渴求。但是過了30歲左右時，因為和老朋友相聚而產生了轉機。兩人開心地談論辻村聰志學生時代熱愛的軍事人物模型改造，而這位朋友委託的製作正是電視劇『必殺仕事人』的主角中村主水。這雖然是忠臣藏1／35比例人物模型的改造委託，卻喚起他創作的樂趣，而進入了造型的世界。此後還承接了山田奉行所紀念館展示用的繪畫，並且和京都一家以博物館展示模型製作為業的企業有了合作機會，承接了國立歷史民俗博物館展示用的250多個人物模型製作。就這樣由中村主水牽起的「造型緣分」，從紀念館展示用的御座船繪畫，到江戶橋往來的250人的人潮重現等博物館展示作品、市售的人物模型改造，再到原創人物模型製作，串起了辻村聰志的造型人生（可惜當時製作的中村主水似乎未留下作品照片，究竟辻村聰志製作的中村主水有怎樣的高品質表現，令人好奇不已）。

During spring break of his first year in Junior high school, Tsujimura's desire to start drawing grew immensely after watching the movie "Nausicaa of the Valley of the Wind." Since then, he continued to paint almost every day for more than 15 years, even after getting a job. Although Tsujimura did not set out to become an artist, he recalls that the joy he felt when people received his works was the prime driving force that kept him drawing Tsujimura got back into modeling after meeting an old friend from his childhood. His friend reminded how great Tsujimura's ability to modify figures was and even proceeded to ask him for a custom scratch-built figure of Nakamura-Shusui, the main character from a famous Japanese television series "Hissatsu Shigotonin." This experience truly introduced Tsujimura to creating figure sculptures and eventually led to various commission works from museums such as the National Museum of Japanese History in Chiba and Yamada-Bugyo Memorial Museum in Mie prefecture.

▶20歲左右描繪的水彩畫作品。辻村聰志自己相當滿意整體構圖，呈現回眸少女美麗的臀部線條。這幅畫充滿變形的技法，藉此稍微強調眼睛等迷人的重點。

Water-color bust of a young girl in a swimming suit. This painting was drawn by Tsujimura in his teenage days.

▶10幾歲後期描繪的水彩畫作品。因為照片人物的臀部線條很美，讓辻村聰志起心動念想要描繪。當時的速度大概是一口氣從傍晚畫到夜晚完成。

Figures can be viewed from various angles, but as with this painting, there is always a dominant angle.

▶約20多歲的水彩畫作品。這也是因為覺得照片中的「光線實在太美」而想動手描繪，花了一整天的時間完成。想將美麗化為有形保留的期望是辻村聰志在繪畫和造型不變的原則。

One must observe, digest, and output information with incredible accuracy to draw in such intricate details.

▶三重縣伊勢市山田奉行所紀念館委託描繪的水彩畫作品「御座船虎丸」，大概是35歲的作品，是尺寸超過20號的巨幅畫。因為御座船虎丸沒有留下圖面，只好依靠文獻資料和研究機關共同完成。

A commission drawing of a traditional Japanese-style boat exhibited in Yamada-Bugyo Memorial Museum.

1/24比例擬真人物模型系列

1/24比例人物模型是長谷川最常推出的商品系列尺寸。雖然原型製作尺寸較小但一點也不輕鬆，因為自己總是「想做得更精細一些」，所以製作時不以為苦，可說是我還蠻偏好的比例。為了讓消費者也能清楚感受這個尺寸呈現的人物表情，自己製作的原則是要精細呈現出眉形和嘴唇賦予人物表情的重點部分。希望大家觀看時能注意到這些地方。1/24比例商品系列中也有以射出成型製成的人物模型，但是從模具完整取出等限制也很多，在姿勢和零件分割方面費了一番功夫。希望方便大家改造成自己喜歡的樣子。

1/24 scale is the most common size in Hasegawa's figure lineup. For me, I have the most fun when putting tiny details into smaller figures, so you could say that 1/24 is my preferred scale. My sculpting policy is to make sure that the critical points of facial expressions, such as the shape of eyebrows and the mouth, are molded perfectly, even at this smaller scale. Typically these figures are cast in resin, but some figures in the 1/24 scale lineup are made with ordinary injection-plastic parts. There are many restrictions with this method, and I always have difficulty figuring out how to divide a figure into each piece. Since plastic is easier to work with than resin, it should be easier to modify these figures to their liking.

LAMBORGHINI MIURA P400 SV
W/ITALIAN GIRL'S FIGURE

藍寶堅尼MIURA P400 SV
w/義大利女性人物模型（2019年）

人物呈邊走邊往旁邊看的動態姿勢，自己曾實際擺
出相同姿勢，多次確認這樣的姿勢是否合理。造型
要留意的地方是需在褶襉重疊的服裝添加稍微飄起
般的線條，呈現輕盈的樣子。我將她的表情設定為
不自覺留意到停駐的車輛，不知道樣子是否傳神。

She's walking past a car but can't get her eye off of it,
it seems. This figure has a lot of momentum in its
posture. I ended up posing similarly and taking
pictures of my self, to make sure she looks as natural
as she can.

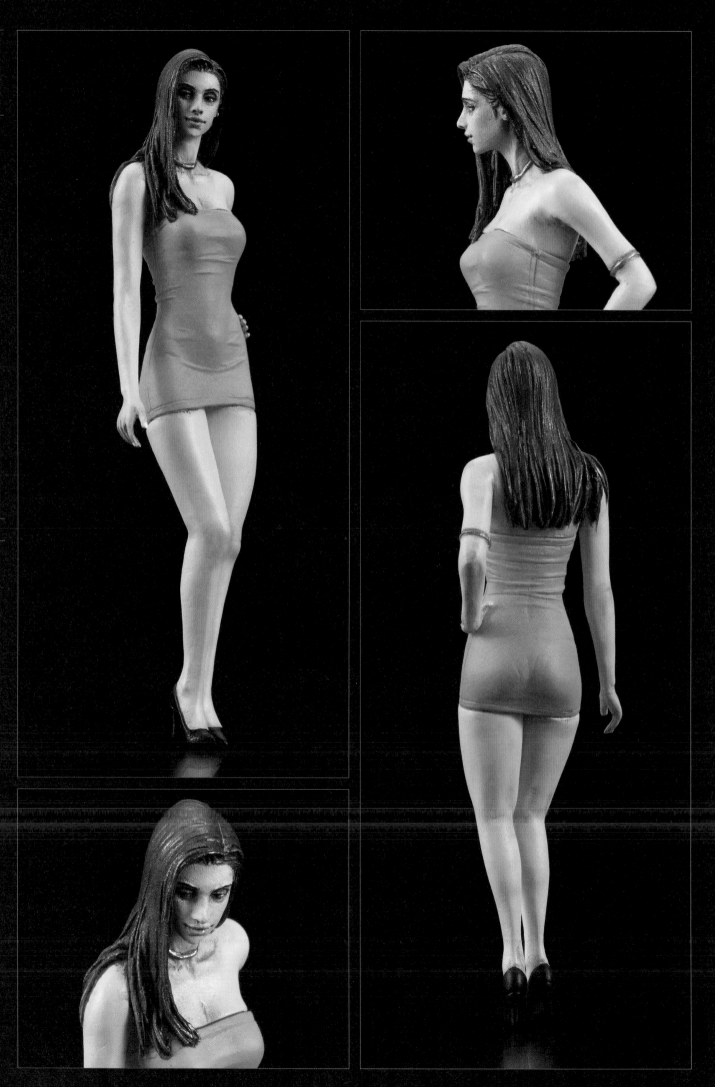

LAMBORGHINI JOTA SVR
W/ITALIAN GIRL'S FIGURE

藍寶堅尼JOTA SVR
w/義大利女性人物模型（2020年）

肌膚稍微偏小麥色，長相偏義大利或西班牙等拉丁系的美女，洋裝款式雖然較暴露但是不會俗氣。這個作品呈現截然不同的風格，整體感覺是一位成熟女性凝視著跑車的樣子。原本的企劃就是藍寶堅尼等義大利車款旁站著一位女性，成品呈現的畫面還不錯。

This is a figure of a beautiful Italian girl with slightly tanned skin. Her dress is revealing, but nothing excessive. The whole premise of the project was to add a figure that looks just right for the Lamborghini, and I think the end-result came out pretty well.

企劃需求不是讓人物站在車輛旁邊，而是希望人物坐在引擎蓋上，並且和車體宛若一體成型。為了讓人物和車體自然貼合，在原型製作時先將人物的接觸面削薄之後添加補土，再按壓在鋪上保鮮膜的車輛套件。我也希望大家注意有別以往的髮型亮點。

1966 AMERICAN COUPE TYPE I
w/BLOND GIRL'S FIGURE

1966年美國轎跑車TYPE I
w/ 金髮女郎人物模型（2019年）

企劃需求不是讓人物站在車輛旁邊，而是希望人物坐在引擎蓋上，並且和車體宛若一體成型。為了讓人物和車體自然貼合，在原型製作時先將人物的接觸面削薄之後添加補土，再按壓在鋪上保鮮膜的車輛套件。我也希望大家注意有別以往的髮型亮點。

A more unified look was required for this figure sitting on the hood of a car. This was achieved but shaving off her butt and part of her legs, adding a thin layer of putty, and squeezing tightly to the vehicle. Excess putty was then removed after the hardening.

1966 AMERICAN COUPE TYPE B w/BLOND GIRLS FIGURE

1966年美國轎跑車TYPE B
w/金髮女郎人物模型（2019年）

姿勢為雙手輕放在車上回眸，強調臀部線條。一邊
看著電影一邊作業，所以帶點女主角的形象。製作
這類作品時也會參考國外雜誌的封面女郎，會有各
種身形可參考。這個女郎屬於身形纖瘦、臀部較小
的類型。

This is a pose that emphasizes her buttocks, looking
back while her hands lightly touching the car. I was
working on this sculpture while watching a movie, so
the film's heroine definitely altered the finish of this
figure.

1966 AMERICAN COUPE TYPE P
w/BLOND GIRL'S FIGURE

1966年美國轎跑車TYPE P
w/金髮女郎人物模型（2019年）

美國轎跑車和女性人物模型組合企劃大受歡迎的第
3 彈商品。最初知道這項企劃時，心想：「長谷川
的決定頗為大膽」。雙腳打開跨坐在車上的姿勢很
容易流於粗俗的感覺，但我覺得自己在這方面做得
挺收斂。聽到銷售方面也獲得不錯的成績，讓我放
心不少。

This was the first installment of Hasegawa's product
which combined American coupé with my figures,
which made me excited and worried at the same time.
I was thrilled to hear that the sales were quite good
with this one!

想像有人坐在實際車輛的引擎蓋上都覺得有些可
怕，雖然如此還是考慮到避免讓高跟鞋刮傷車子，
而調整了腳尖和腳踝的角度。側坐的姿勢雖然沒有
想像中困難，但是這次放在腳上的左手還需要處理
零件分割，這個部分的難度則超乎想像，讓我吃盡
苦頭。

1966 AMERICAN COUPE TYPE C
w/BLOND GIRLS FIGURE

1966年美國轎跑車TYPE C
w/金髮女郎人物模型（2019年）

想像有人坐在實際車輛的引擎蓋上都覺得有些可
怕，雖然如此還是考慮到避免讓高跟鞋刮傷車子，
而調整了腳尖和腳踝的角度。側坐的姿勢雖然沒有
想像中困難，但是這次放在腳上的左手還需要處理
零件分割，這個部分的難度則超乎想像，讓我吃盡
苦頭。

Thinking about it, riding on the hood of a car is a
scary and very unnatural situation. Even so, I took
care to adjust the angle of her heels to avoid
damaging the car's body. I remember struggling a lot
with the way of dividing each part.

1966 AMERICAN COUPE TYPE T
w/BLOND GIRL'S FIGURE

1966年美國轎跑車TYPE T
w/金髮女郎人物模型（2019年）

製作時將人物設定為拉丁系女孩，稍微曬過的健康
膚色，搭配海灘或夏季慶典都很適合，讓人有擺脫
日常釋放壓力的感覺。回眸的姿勢可以突顯臀部的
線條，作品表現上充滿挑戰。因為是站立的姿勢，
或許和車輛的成套感稍嫌薄弱，但搭配車輛以外的
小物或許也會有不錯的效果。

"A Latin girl with a slightly tanned and healthy-looking
skin tone who is ready to hit the beach or have a blast
in a summer festival" was the motif I had for this
figure. It might be a good idea to combine her with
accessories other than cars.

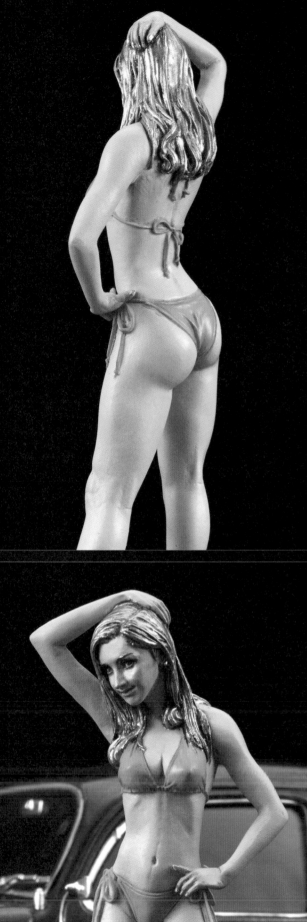

VOLKSWAGEN BEETLE TYPE 1 (1966) "CAL LOOK"
w/BLOND GIRL'S FIGURE

福斯金龜車（1966年）「露營拖車」
w/金髮女郎人物模型（2020年）

脖子微傾，眼神往上看向對方，表情風情萬種。大
腿的造型做得有點結實，所以臀部也稍微做得豐滿
一些。想做成類似國外訂製車雜誌開卷拍攝的封面
女郎。手撥頭髮的樣子也是這類雜誌經常出現的姿
勢，這或許是一種經典姿勢。

This one looks like it's participating in a photoshoot
for the front cover of a magazine. She has her head
slightly tilted and her hands scooping up her blonde
hair. Pose such as this is frequently spotted in such
entertainment magazines.

VOLKSWAGEN TYPE 2
DELIVERY VAN "FIRE PATTERN"
W/BLOND GIRL'S FIGURE

福斯金龜車 TYPE 2
廂型車「火焰造型」
w/金髮女郎人物模型（2020年）

她看來就是一副馬上要去海水浴場的樣子。乍看之
下腰間骨盆似乎頗有分量，但我覺得這樣的身形比
例較為協調。臀部的造型重點在於形狀和分量會隨
著泳裝設計而有所不同。另外，泳裝形成的皺褶也
會因為設計和材質而有各種樣貌。

She looks ready to go swimming! I always make sure
to adjust the bodyline of each figure according to the
costume their wearing. So her hips might look a little
big at first glance, but it's actually perfect for this style
of swimsuit.

1/4 TON 4X4 UTILITY TRUCK
w/BLOND GIRL'S FIGURE

1/4噸 4×4軍用卡車
w/金髮女郎人物模型（2020年）

比起現在，造型給人稍微有點年代感，例如側邊設計稍微較寬的舊款泳裝和眼影較濃的妝容。白色邊緣有點聖誕節的感覺。人物坐在引擎蓋，腳放在輪胎的姿勢，腳彎曲的方式、身體和車輛的貼合度等都是製作上較費心的地方。

This girl is fancying a slightly older style of swimming suit with its sides wider than most modern swimsuits. White borders on her swimming suit give a Christmas vibe. It took a lot of trial and error to get it to fit perfectly with the Jeep.

Pkw.K1 KÜBELWAGEN TYPE 82 w/BLOND GIRL'S FIGURE

Pkw.K1桶車82型
w/金髮女郎人物模型（2020年）

髮色和捲度都偏復古的髮型、外套下襬打結使胸部
顯得更豐滿等都是造型亮點。腳因為零件分割的關
係其實做到裙子內側。軍帽等小物都是參考自己原
本的收藏，成了有助製作的資料。考慮貼合度而將
頭和帽子做成同一個零件。

The highlight of this figure is the slightly old-fashioned
hairstyle, and her breasts held tight with her jacket.
Although not seen, the inside of her skirt is sculpted
too. As for her cap and other accessories, my military
collections were used as a reference.

1/4 TON 4X4 UTILITY TRUCK
(CAL. 50 M2 MACHINE GUN)
w/BLOND GIRL'S FIGURE

1/4噸 4 × 4軍用卡車（50口徑M2機槍配備）
w/金髮女郎人物模型（2021年）

由於一開始收到的草稿畫就非常迷人，所以直到最後都依照原先的方向製作。為了呈現臀部翹起、手放在槍枝這樣散發魅力的姿勢，一定會從設計較短的裙子露出一半以上的臀部，但是長谷川表示由我全權負責裸露的程度，所以就決定這樣的設計。

Her hips sticking out, her hands on the gun, it's a typical glamour pose we're all used to seeing. I had full authority on determining the amount of her skin exposure, so her breasts and butt ended up being mostly exposed.

Pkw.K1 KÜBELWAGEN TYPE 82 (BALLOON TIRE) w/BLONDE GIRL'S FIGURE

Pkw.K1桶車82型（充氣輪胎）
w/金髮女郎人物模型（2020年）

嚴肅的表情與其說是軍人的樣子，倒不如說是想營造有軍人氣般的女性。頭戴的船形帽和軍裝用品除了參考自己原先的收藏，自己對於實物也有一定的理解，所以將帽子設計成正確的戴法。沒想到自己對於軍事用品感到興趣，成了製作這類作品的助力。

Her crisp expression might seem soldier-like, but it's more of a typical expression of a female with confidence. I'm also a big fan of military subjects, so my knowledge and resources came in handy for this

DATSUN BLUEBIRD 1600 SSS
w/60'S GIRL'S FIGURE

達特桑藍鳥1600 SSS
w/60年代女性人物模型（2021年）

製作時特別留意昭和40年代女性的服裝，還有臉部
大小等身形比例。我有特地參考當時女性的照片等
資料，歸納出的特點是肩寬和腰圍都比現代女性窄
瘦。整個作品的故事背景是現在70多歲的老婆婆回
憶舊時過往的樣子。

I sculpted her retro costume while paying attention to
the size of her face and the balance of the whole
body. You may notice that her waist and shoulder
width are slightly slimmer than the more modern
female figures I've made in the past.

SUBARU 360 YOUNG-SS
w/60'S GIRL'S FIGURE

速霸陸360 YOUNG SS
w/60年代女性人物模型（2021年）

這個人物設定在昭和30年代，我收到當成參考資料的當時照片甚至大多都是黑白照。不但服裝設計充滿復古感，髮型也很有當時的風格。故事設定為和男友約會途中的抓拍照片，女生假裝自己是車展的展場女郎，擺出拍照的姿勢。

The design of her attire has a retro feel, and the hairstyle is also reminiscent of the period. I decided to pose her as if she is pretending to be a companion at a car show, being photographed by her boyfriend on a lovely summer afternoon.

TOYOTA 2000GT
w/60'S GIRL'S FIGURE

豐田2000GT
w/60年代女性人物模型（2017年）

因為是汽車模型搭配人物模型的企劃第一彈，所以在製作造型時最重視的是兩者成套一組的感覺。大鈕扣設計的服裝是當年最時髦的造型，但其實人物的臉和風格都做成通用的設計，這是考慮到方便讓人物模型也可以搭配汽車以外的物件。

Since this was the first product that combined my figures with Hasegawa's automobile plastic models, I did my best to unify the two, the girls and the car, as naturally as I can. Large buttons on her dress were the latest fashion feature in the 1960s.

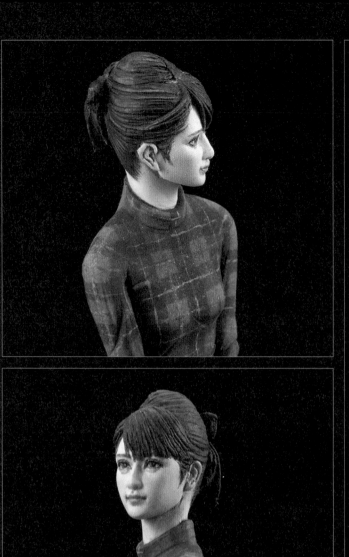

MAZDA COSMO SPORT L10B
w/60'S GIRL'S FIGURE

馬自達COSMO SPORT L10B
w/60年代女性人物模型（2017年）

主題是昭和40年代當時的千金大小姐，重現了當年
日本人的身形。若考慮當時的時代背景，大約是戰
後20年的時候，從資料照片也可看出當時女性大多
身形纖瘦。衣服的格紋圖案當然是用手繪完成，因
為實在有太多人提出「自己也想畫出相同圖案」的
要求，我甚至還在twitter上傳了塗裝方法的訣竅。

The theme was to recreate a young Japanese girl from
the 1960s. This was about 20 years after the war,
and we can see from period photos that there were
not many women with busty bodies. The checkered
patterns on the dress were hand-painted.

HOW TO MAKE BEAUTY

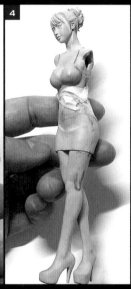

1 基本素體，到這個階段已經過多次的反覆修正。身體曲線的造型幾乎都是用添加補土的塑造手法完成。

2 確認素體造型設定的零件分割。紅色線條標示的地方是洋裝的外輪廓線，雙臂的分割為暫定狀態。

3 確認裙子內側、腳的分割線條和裙子部分的嵌合。在確認了最終造型後會開始分割成零件，所以要再重新製作嵌合的部分。

4 製作裙子等衣服造型時，會覆蓋上擀得極薄的環氧樹脂補土。

5 追加裙子皺褶等基本細節後的狀態，並大概做出捲髮的細節。

6 考慮塗裝的方便，手臂沿著洋裝肩帶分割。脖子的首飾和洋裝的裝飾都用擀薄的環氧樹脂補土製成，再用刻紋表現縫線。

7 將近完成的階段。零件接合也沒有問題的完成狀態。

1. The base of the body before adding details. It takes a lot of trial and error to get to this point.

2. Posture of the figure is finalized, and some sections are sawed off, dividing them into parts. Keep in mind that the attachment points for both arms are still in very rough shapes.

3. Examining the fit and assembly method of the skirt. 1/12 scale figures can get quite heavy, so they are typically made hollow in the middle to reduce strain and improve handling.

4. Thin sections such as the skirt and other sections of the clothing are created by wrapping thinly rolled epoxy putty around the base body.

5. Basic details such as creases of the dress are added. Her hair has received some more detail.

6. For easier painting, both arms are divided at the shoulder straps of the dress. Her fancy neckless and finder details of the dress have also been added with thinly rolled epoxy putty.

7. The upper body of the nearly completed figure. The fitting of each part is excellent and seamless.

▲辻村聰志在臉部造型方面最注重 3 點，分別是耳朵高低、眼尾構造，還有讓臉頰呈現漂亮的 S 形弧線。可以利用光線反射確認臉頰線條。

There are three key elements when sculpting a female figure's face: the height of the ears, the line of cheeks, and the outer

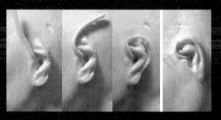

▲耳朵是表現一個人個性的重要部位。先用補土做出大概的形狀並等其硬化，再添加上分成細條狀並圍成圈的補土並等其硬化。最後嵌入間隙中並且用沾濕的筆尖調整形狀。

Ears are an essential part of a face. After creating a rough shape of an ear with a lump of putty, additional thin slices were added to create depth and blended with a brush.

▲衣服的細節用擀成薄片狀的環氧樹脂補土塑形。擀薄補土時會使用擀麵棒和太白粉。裙子利用添加補土和雕刻作出形狀。

Details of the clothing were sculpted using epoxy putty stretched into thin sheets. You can avoid putty from sticking to your roller by using powder such as starch or flower.

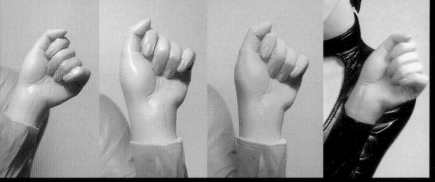

◄底塗完成後，在指尖等陰影部分塗上基本膚色前，先用酒紅色調的琺瑯塗料入墨線。接著在表面用噴筆噴上一層薄薄的肌膚基本色，這樣連內側都會呈現出自然的陰影，比起肌膚塗裝完成後再入墨線，這個做法不但不容易產生色塊，成品也較為漂亮。另外，指尖、手掌為粉紅色，手腕內側則噴上薄薄的白色，這樣即便沒有塗上代表靜脈的藍色，都能有效呈現擬真質感。

Areas between the fingers were painted with wine-red enamel paint. Then the entire hand was coated with base skin tone using an airbrush. This method allows you to achieve a more natural gradation between highlights, base color, and shadows.

◄▼眼睛塗裝使用琺瑯塗料。先描繪眼白，再用焦茶色描繪睫毛和眼睛輪廓。眼睛大概有三層色調，先用中間色塗在眼睛輪廓內，接著開始重疊塗出眼睛明亮處和陰暗處。這時分別斜向塗色就能呈現有深度的感覺。斜線的上側為眼睛的陰暗色，下側為明亮色。若是茶色眼睛，明亮部分為粉紅色，陰暗部分為藍色調的紫色。這裡不使用以茶色系為基底調出的顏色，是為了避免顏色顯髒。眼睛中心的瞳孔使用黑色，畫圓且左右大小要畫得一致。陰暗部分則用白色表現打亮，先用消光漆為整體鍍膜，最後再用清漆添加光澤。

Enamel paints are used for painting eyes. First, the whites of the eyes are painted, followed by the eyelashes and the outline. Pupils are painted with three basic tones, after which the highlights/shadows are added. Finally, both eyes receive a clear gloss coat.

▶在呈現金屬質感的部分，例如洋裝肩帶和首飾等地方，一一用筆塗仔細塗上硝基漆的金屬塗料。另外，建議衣服和肌膚、頭髮和肌膚等交界，使用紅藍色混合的棕色調琺瑯塗料添加入墨線。這樣就不會使顏色交界產生暈染的情況，讓整體呈現清晰的輪廓線條。這次解說的是1/12比例的完整工序，但是基本上1/20比例和1/24比例的模型也是類似的工序流程。

Straps of the dress and the neckless were painted using metallic lacquer paint. Thinly diluted enamel paint (brownish red/blue) was applied at the edge of her hair and skin for a smoother transition of color. Same techniques can be applied to 1/20 and 1/24 scale figures as well.

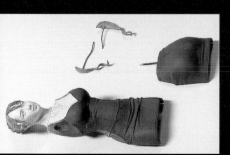

▲肌膚或衣服等基本塗裝都使用硝基漆塗料並且用噴筆塗色。配合沿著身體曲線的起伏和皺褶塗上漸層色調，細節處則使用琺瑯塗料筆塗上色。使用硝基漆塗料是因為如果塗裝失敗較方便重塗。

The base color of the skin and the dress were painted with lacquer color using an airbrush. For finer details, brush painting with enamel colors is preferred.

80's BUBBLY GIRLS FIGURE

80年代拜金女郎人物模型（2018年）

人物模型主要為塑膠套件，由於是我第一次負責這類原型的製作，所以這件作品讓我印象深刻。原型製作的工序基本上並沒有不同，但是配合模具的修改作業相當繁多。其實我在泡沫經濟時代時還是個學生，我對於當時的時尚流行印象薄弱，不過這倒讓我在製作時充滿新鮮感。

This was the first time I sculpted a figure which was to be cast in injection-plastic, so it was a very memorable project. I was still a child during the 1980s, so working on their costume was a fresh and exciting experience.

90's PLATFORM BOOTS GIRLS FIGURE

90年代厚底鞋辣妹人物模型（2018年）

因為是和自己同時代的女性，所以印象深刻。從身形比例可以看出不同於昭和時代的女性，刻意表現出腰部纖細和注重妝容的氛圍。眉毛也很細。考慮到模具成型的狀況，手提包包的背帶和手拿手機的手指姿勢都得費心設計。

Unlike the 1980s, the 1990s was the generation I'm most fond of. Girls from this era had their proportions dramatically changed from the women of the Showa period. Their waits were narrower, which were reflected accordingly to these figures.

90's PADDOCK GIRLS FIGURE

90年代賽車女郎人物模型（2018年）

其實我沒有親眼看過賽車女郎（笑）。作品主題設定在80年代末到90年代初，正逢日本掀起賽車風潮之時。兩人都是日本人，但是身形稍有差異。陽傘是由長谷川以數位設計提供。

To be honest, I have never seen a race queen face to face. The motif of these girls is the late 1980s to the beginning of the 1990s, just when motorsport hit its prime in Japan. The parasol was digitally designed by Hasegawa.

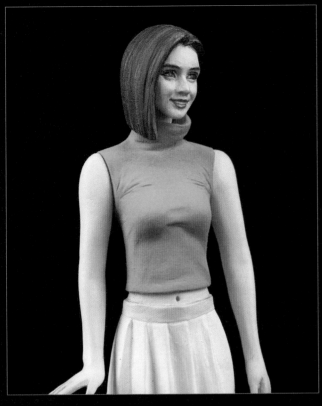

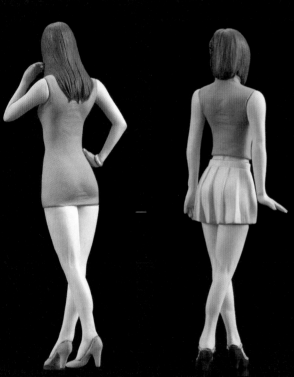

FASHION MODEL GIRLS FIGURE

時尚模特兒女性人物模型（2019年）

塑膠製人物模型至此已來到第 4 彈，但我依舊尚未完全熟悉模具需求，原型製作時依舊過著艱苦奮鬥的日子。我很在意纖瘦身形，以及衣服是否貼合身體的表現。尤其重現高領衣服會有模具的限制，難度頗高，但是辛苦獲得回報，成品呈現不錯的效果。

This is the fourth injection-plastic figure Ive created, but I was far from feeling comfortable with the technical difficulties behind it. It was especially difficult to recreate the green high-neck costume due to the limitations of the mold, but the hard work paid off!

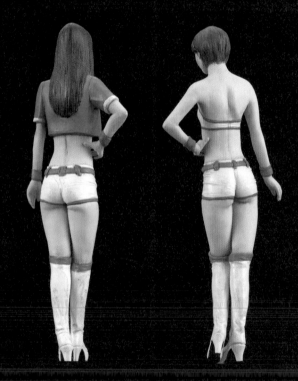

COMPANION GIRLS FIGURE

展場女郎人物模型（2019年）

除了緊實的身形，還留意要呈現良好的身材比例。
為了讓人物顯高，在比例上特別將臉部做得小一
些。短髮人物的臉部希望做成既像日本人又像外國
人的感覺，不過看成現代女性或許較為自然。這個
組合除了可以運用在展場女郎也可以當成賽車女
郎。

I would say these girls are the top-rated companions
with their very good proportions and tight body. Their
faces are sculpted slightly small in an attempt to make
them look taller. The girl with the shortcut could be a
foreigner, or just modern-looking Japanese.

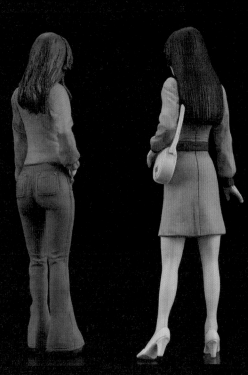

70's GIRLS FIGURE

70年代女性人物模型（2019年）

我並沒有特別尋找參考的模特兒，不過我記得銷售時常聽到有人反應「好像○○」。這個時代流行的輝夜姬髮型和衣服剪裁相當有特色。只在連續劇中看過的喇叭褲在質感和比例的拿捏相當困難。

The 1970s was an exciting period in terms of fashion, with distinctive hairstyles such as "Princess Kaguya" and clothing in vivid color. I've only seen pantaloons in TV dramas, so getting the appropriate texture and detail was challenging.

HOLLYWOOD CELEBRITY GIRLS FIGURE

好萊塢名人女性人物模型（2019年）

雖然是看著實際資料設計和決定服裝的設定，但是除了分割位置的例行問題之外，最在意的是整體氛圍是否呈現出「名人穿搭的高品牌質感」。在克服了重現服裝細節的難題後，一如往常地又得面對包包背帶的零件分割問題。

We arranged and decided on the design of the costumes while looking at actual pieces of clothing. Usual problems such as "how to divide parts" persisted, but we were more concerned about capturing the "celebrity" vibe accurately.

80's GIRLS FIGURE

80年代女性人物模型（2020年）

從松田聖子的髮型就可以看出，在這個時代當紅偶像的形象對大眾流行有多大的影響力。雙手在前面交疊的姿勢也刻劃出當時的偶像文化。髮型不論前後左右都相當有分量，所以為了重現這些細節思考了各種分割方式。

In this era, we can see how the image of the top idols of the time, such as Seiko Matsuda, influenced the fashion of the general public. The pose of her two hands folded together in front is another way of capturing the idol culture of the 1980s.

PADDOCK GIRLS FIGURE

賽車女郎人物模型（2020年）

造型大約是2010年前後的人物形象，和90年代的賽車女郎相比，腿長臉小，比例和表情完全不同。另外，髮型設計也變得較為複雜，所以分割難度變得更高。零件的嵌合度就是無法如樹脂鑄造那樣完美，讓我心中留有遺憾。

Compared to the race queens of the 1990s, their legs are longer, and faces are smaller as if they are separate species. Also, their hairstyles have become more complicated, giving me extra challenge sculpting and detailing.

50's AMERICAN GIRLS FIGURE

50年代美國女性人物模型（2021年）

臉部造型特意擷取好萊塢女星的優點，我自己很滿意。我還努力重現可愛的髮型。大蓬裙遮蓋了身形，而我總是從未著裝的素體開始造型，所以反倒是有難度的一個主題。參考資料也是只有70年前的黑白照片，讓我印象深刻。

I think I accurately represented the facial features of typical Hollywood actresses in the 1950s with these figures. These were one of the most challenging projects I worked on since their body lines are mostly hidden with fluffy skirts.

其他擬真人物模型

1/20比例模型是我個人製作原創作品時習慣的尺寸，所以製作起來非常得心應手。比起1/24比例的模型，全長大概只高了1cm左右，製作起來輕鬆許多。製作自衛隊軍官系列時收集了大量的資料，我非常注意手敬禮的角度和姿勢的正確。機鼻藝術女郎因為原作是插畫，所以是將插畫變成寫實的作業。因為有不少人都看過原本的插畫，所以製作時會特別留意不要和原作差異太大。超人力霸王系列作品中，都是由女星擔任模特兒，所以桌上擺放的資料比過往要多很多，臉部造型也格外花時間。

1/20 used to be my go-to scale before collaborating with Hasegawa. Generally speaking, 1/20 scale figures are about a centimeter taller than 1/24, but that centimeter makes all the difference. For the "Self-Defense Force" lineup, I gathered as many resources as possible to get the correct stance and angle of salute. "Nose Art Girls" are based on famous nose art illustrations of various aircraft, so I took extra care to capture the original artwork's touch. Finally, figures belonging to the "Ultraman" lineup were, of course, sculpted with the real-life actress in mind who played each character. I remember my desk being messy with all the resources I collected and sculpting their faces took more time than anticipated.

F-15J EAGLE
w/J.A.S.D.F. FEMALE PILOT FIGURE

F-15J 鷹式戰鬥機
w/航空自衛隊女性飛行員人物模型（2020年）

現在雖然有航空自衛隊女性飛行員，但是收到的委託是虛構的女性飛行員。我聽說女性自衛隊軍官大多將頭髮綁起或是短髮造型，造型做成髮型俐落的20多歲幹部。造型設計雖然簡單，但是服裝皺褶等起伏多變，希望帶給大家塗裝的樂趣。

Although actual female pilots exist within Japan Self Defence Force today, this figure was ordered as a fictional female character. I've heard somewhere that most female JSDF officers wear their hair tied up or simply cut short.

J.M.S.D.F. DDG MYOKO
W/FEMALE S.D.F. OFFICIAL FIGURE

海上自衛隊妙高號護衛艦
w/自衛隊女軍官人物模型（2020年）

主題縮臂敬禮是在船艦特有的場景，講究指尖輕觸
帽簷的位置和指尖微開的角度都是自衛隊獨特的細
節。服裝顏色較深，所以縫線確認困難，這裡我想
呈現背部打直、端正的敬禮姿勢。

The position of the fingertips, lightly touching the cap's
brim, and the angle at which the toes open are details
unique to the Self Defence Force personnel. I think I
was able to recreate an elegant and accurate salute
or a soldier.

OH-6D "AKENO SPECIAL 2019"
w/FEMALE S.D.F. OFFICIAL FIGURE

OH-6D「明野SPECIAL 2019」
w/自衛隊女軍官人物模型（2020年）

製作的人物模型恰好是在自家附近陸上自衛隊駐軍
基地負責飛機的自衛軍官，讓我有種微妙的親切
感。表情較為輕鬆並且帶著微笑，但依舊散發正氣
凜然的氣質。制服塗成迷彩色使縫線和口袋等細節
變得難以辨識，不過做得相當精細。

I felt a strange sense of familiarity with this figure of
the Self Defence Force officer attached to a garrison
near my hometown, Akeno. She is in a relaxed
position, smiling lightly but with a dignified expression
and posture.

NOSE ART GIRL'S FIGURE
"LEROY'S JOY"

機鼻藝術女郎人物模型
「LEROY'S JOY」（2020年）

原本只是像草稿般的機鼻藝術，所以要將其轉換成
寫實的造型，我覺得是很具挑戰性的企劃但也樂在
其中。形象為五官立體分明的拉丁系女性，但整體
設計稍微做了一點改變，並留意不要太偏離大家印
象中的機鼻藝術。

It was a challenging project to translate historical nose
art into a three-dimensional figure. Yet, I remember
enjoying the process a lot. I added some originality to
the figure while being careful not to depart too much
from the original artwork.

NOSE ART GIRL'S FIGURE "BLONDIE"

機鼻藝術女郎人物模型
「BLONDIE」（2020年）

這也是從插畫轉為寫實風格的人物模型，瀏海分量較多，形狀的掌握上也有些難度，側躺姿勢的接觸面究竟該如何設計，這些從2D轉換成3D的差異調整，讓我費了一番苦功。相反的可以自由發揮的部分也很多，帶來很多創作上的樂趣。例如設計了背部裸露面積較大的洋裝。

This one was also from the famous nose art "Blondie" from the 4th Fighter Group. Her hairstyle was an absolute nightmare to represent accurately, but on the other hand, I had a lot of freedom and fun while working on this one.

JET VTOL
w/AKIKO FIGURE

JET VTOL
w/AKIKO隊員人物模型（2019年）

其實我自己並不熟悉特攝作品，在作業時和平常製
作原型一樣，先確認主題圖像並不斷模擬。除了長
谷川提供的資料之外，自己手邊也收集了一些資
料。由於自己並不曉解超人力霸王的作品，所以作
業時分外感到新鮮有趣。

To tell the truth, I'm not very familiar with special-effect
TV series such as "Ultraman." But just as usual, I
referenced tons of images of the motif and devoted
myself to the task of accurately representing her face
as well as the costume.

©圓谷PRO

現在雖然不太使用SNS，但是聽說這個人物模型在
銷售時，真的很多作品的粉絲給予不錯的評價。
安全帽和槍套等細節的努力讓我充滿成就感。平常
不會特別處理，但是為了讓這件作品給人更迷人的
感覺，稍微調整了臉部造型的比例。

MAT-VEHICLE
w/MAT FEMALE CREW

MAT VEHICLE
w/MAT女性隊員（2018年）

現在雖然不太使用SNS，但是聽說這個人物模型在
銷售時，真的很多作品的粉絲給予不錯的評價。
安全帽和槍套等細節的努力讓我充滿成就感。平常
不會特別處理，但是為了讓這件作品給人更迷人的
感覺，稍微調整了臉部造型的比例。

I hardly use any social networking services now, but I heard
that this figure was received very positively amongst
modelers and fans of the TV show. I usually don't do this, but
I slightly simplified her facial features to make her more
charming.

©圓谷PRO

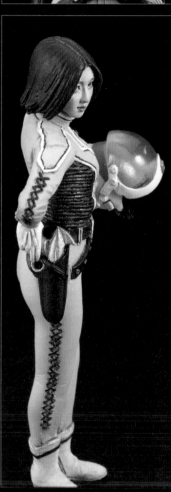

ULTRA HAWK 001
w/ANNE FIGURE

ULTRA HAWK 001
w/ANNE隊員人物模型（2018年）

ANNE隊員的衣服資訊量極大，對於手工製作原型的我來說是很繁重的作業。側邊的蛇腹狀、編織交叉和安全帽都很複雜。這些也都有很多資料可以參考，不過髮型會因為劇中時期而有所變化，為了呈現相似的長相對於主題的選擇稍微有點苦惱。

This one has a lot of detail to work on, especially for her costume. The side bellows, the braided cross, the helmet, to name a few. Her hairstyle changes several times during the show, so I had difficulty deciding which to pick.

©圓谷PRO

長谷川是如何開始銷售
由辻村製作原型的人物模型

The behind the scenes of Tsujimura x Hasegawa colaboratory figures

訪問與撰文／寒河江雅樹

提到長谷川或許大多數人的印象是這間是走寫實路線的汽車模型廠商，最近全力投入汽車模型發展。當然身為一家老牌模型廠商，從推出的商品系列可以感受其企劃風格多元的靈活策略，例如：連搭乘戰鬥機的飛行員、操控賽車的駕駛等周邊品項都做成精良的商品，而且在 1990 年之後，還推出了遊戲或動漫作品中的機械造型或角色人物等系列商品。話雖如此，但誰能料想到由辻村聰志製作原型的人物模型竟會為這些商品系列帶來如此效益。究竟是甚麼原因長谷川選擇和辻村聰志這位原型師合作？我決定訪問長谷川企劃負責人，我想了解在他很快發覺到辻村聰志造型的魅力，並且運用在主要商品系列時，是如何啟動擬真女性人物模型的企劃和當時的過程。

「2014年我們推出的商品『擊墜王－蒼空 7 人－（第2次世界大戰王牌飛行員和 7 架戰機套組）』，是以1/48比例製作多架戰機和擊墜王人物模型的組合，由此開始了我們和辻村先生的合作。商品中有一名蘇聯的女性王牌飛行員『莉狄亞·利特維亞克』，成了我們公司第一次製作的擬真女性人物模型」。（長谷川商品企劃負責人，以下簡稱長谷川）

這次的商品有附上第 2 次世界大戰各國的王牌飛行員，這些人物模型和飛機一樣都是1/48比例的樹脂製品。人物立像的全高只有 3cm左右，不過表情生動清晰可見，人物模型的細節令人驚嘆，蔚為話題。這時辻村聰志負責 7 位擊墜王中 5 位的原型製作，而其實還包括了當成贈品人物模型的「莉狄亞·利特維亞克」。

「我們在開發『擊墜王－蒼空 7 人』商品的當時，辻村先生早已相當知名，但是我們尚未與他直接聯絡有關原型師方面的工作。正當我們不斷尋找適合的人選，希望有人能夠同時將男性王牌飛行員以及女性飛行員莉狄亞·利特維亞克都做成有魅力的模型時，公司外部的模型相關人員向我們介紹了辻村先生。」（長谷川）

不過故事自此開始突然有很大的發展。在長谷川隔年2015年推出的建築機械塑膠模型，也搭配了由辻村聰志製作原型的人物模型。這一方面歸功於莉狄亞·利特維亞克模型令人驚嘆的擬真，一方面是因為在銷售1/35比例的『雙臂挖掘機ASTACO NEO』時，在套件中搭配女性操作員的想法。不同於前面提到的飛行員人物模型，這次的人物模型和建築機械本體一樣都是塑膠射出成型製成，也就都是塑膠模型。這種建築機械搭配女性操作員的成套商品受到大眾歡迎並且符合市場喜好，之後納入商品系列製作的建築機械都以相同模式銷售。然而從前面的敘述可知，長谷川一開始並未將擬真女性人物模型做成銷售的商品，而只是銷售商品的配件人物模型。那麼長谷川是從哪一個品項開始將擬真人物模型當成主要商品吸引消費者購買呢？

「在推出『雙臂挖掘機ASTACO NEO』之後，營業部提出要銷售1/24比例汽車模型搭配女性樹脂人物模型的成套商品，而推出的商品正是『豐田2000GT w/60年代女性人物模型（2017年11月銷售）』和『馬自達COSMO SPORT L10B w/60年代女性人物模型』（2017年12月銷售）。這兩項商品並非測試市場的實驗性商品，而是如今擬真人物模型商品的初始商品。由於這兩項商品頗受市場歡迎，營業部提議將搭配汽車模型的女性人物模型做成塑膠模型並且銷售，而推出了『80年代拜金女郎人物模型』（2018年6月銷售）。這個商品銷售成績極佳，我覺得這是長谷川擬真女性人物模型廣為人知的契機。從此以後擬真人物模型就成了長期銷售的商品。」（長谷川）

不知道是否有注意到一點？在樹脂製女性人物模型和豐田2000GT成套銷售後僅半年的時間，就開始銷售射出成型製成的拜金女郎塑膠模型。即便之前在建築機械模型中的女性人物模型受到市場歡迎，但是發展至推出商品的速度之快令人相當吃驚。這可說是一種證明，真實反映出女性人物模型受到高度歡迎的程度。像長谷川如此規模的廠商當然會考量模具預算等年度計畫費用，然而從前面敘述的故事來看，這並非例行計畫的一部分。如果我們回推原型製作委託到模具開發、銷售的期間，這段過程發展的速度真的快得令人難以置信。而且也不難想像這項決定影響甚鉅，足以撼動之後的商品系列和長谷川的整體方針。

「『80年代拜金女郎人物模型』在市場的反饋極佳。從那時開始到現在，關於擬真女性人物模型的企劃，公司大多會採納營業部的意見，並且在重視市場銷售趨勢和反饋下推出商品企劃。」（長谷川）

我們了解了這是長谷川正式開啟擬真女性人物模型銷售的轉折點，那麼長谷川和辻村聰志攜手創造女性人物模型的過程又是如何呢？尤其從1/12比例發展的

▲在製作『擊墜王－蒼空7人－』附屬人物模型的原型之後，辻村聰志又製作了1/32比例的尤蒂萊南。此為附屬於2016年4月銷售、1/32比例的Bf109G-6。這時辻村聰志盡情發揮了至今個人持續製作軍事人物模型累積的技巧和實力。

After working on the ace pilot figures, Tsujimura proceeded on sculpting famous Finnish pilot Edward Juutilainen, which was released in April 2016 along with 1/32 Bf109.

▲在尤蒂萊南銷售後僅2個月的時間，2016年6月銷售了「川西N1K2-J局地戰鬥機紫電改『紫電改的真紀』」（1/32比例）將『紫電改的真紀』的主角羽衣真紀做成1/32比例的立體原型。由於一個擬真人物模型作品到這個作品的風格轉變，辻村聰志向大家宣告了自己也可以製作出動漫作品的1/32比例人物模型原型。

Based off a popular comic series, figure of Maki Hagoromo was included in a special edition of Hasegawa's 1/32 N1K2-J Shiden-Kai.

▲2017年2月銷售的「戶外越蛋機少女系列No.01 Amy McDonnell」。在正式開始製作擬真女性人物模型之前，辻村聰志就已製作了動漫人物模型的原型。不過這無疑為截至2021年長谷川「人物模型商品系列」打下了基礎。

Before realistic figures, girls with animated touch were the prime trend for Tsujimura and Hasegawa's collaboratory product lineups.

『1/12比例擬真人物模型系列』等，系列當初從女騎士開始了商品系列，消費者可以理解為「這是為了搭配1/12比例的摩托車所製作的人物模型」，但是之後卻推出了身穿泳衣的封面女郎等商品系列，消費者開始感到困惑：「這是要搭配哪一種機械模型呢？」由此想了解的是人物模型的主題是由長谷川還是辻村聰志設定？另外，還想了解至今銷售最好的商品是哪一個品項？雖然有點八卦，但大家對此都很感興趣。

「人物模型的題材和主題都是由我們公司提出。雖然人物設定是日本人或是外國人沒有太明顯的差異，不過我覺得具備話題性和主題明確、容易了解的商品最熱銷。銷售最好的是1/24比例人物模型系列『80年代拜金女郎人物模型』，在當時引起大家熱烈討論，初次銷售的數量相當多。其他同系列的『賽車女郎』（商品編號：FC03、FC09）等雖然沒有如此熱賣，但很容易找到可搭配的模型場景，屬於長銷型的商品。另外，這些商品都屬於價格較實惠的塑膠模型，而附樹脂人物模型的套件中，1/24比例的『MAT VEHICLE w/MAT女性隊員』銷售量最多。1/12比例的擬真人物模型則是金髮女郎Vol.4，也就是兔女郎的銷售量最多。」（長谷川）

的確現在長谷川推出的汽車模型大多為80年代表性的日本國產跑車，另外再次製作的汽車套件，也都是80年代到90年代參加日本國內外賽車比賽的賽車，非常受到市場的歡迎。可以說前面提到的人物模型的確最適合擺放在這些汽車模型旁邊。由此也可看出賽車女郎等也很適合當作改造的素體。另外，從實際車款受歡迎的程度證明了大家對MAT VEHICLE的喜愛，而且在後來推出的商品中，除了「ULTRA HAWK 001」套件（機體搭配特攝少年心中永遠的女主角）之外，其他商品的銷售數量恐怕都無法與之匹敵。兔女郎我想不需多說一定是會受大家歡迎的商品。

接下來想請問，從原型製作委託到交件的過程，也就是人物模型實際完成的作業流程。

「如前面所說，決定題材和主題後，我們會彙整好姿勢、服裝和細節資料才委託辻村先生。姿勢會由我們公司先描繪出草稿。進入原型作業後，我們會在添加服裝等造型前的素體狀態起分階段確認原型。主要針對細節和整體比例確認在生產時是否會產生問題，例如：若以樹脂鑄造製作時，會確認從矽膠模具脫模的情況，若以射出成型製作時，會確認從模具脫模時是否會有卡住的情況。基本上我們會透過電子郵件傳送照片確認彼此的共識，最後再確認實際成品，如果都沒有問題則交由模具廠商複製。從原型製作委託到交件，雖然每個品項稍有不同，但是大約都落在3～4週之間。收到複製品之後再交給辻村先生塗裝樣本，這就是大致的流程。」（長谷川）

這樣說或許不是很恰當，不過通常大多數的廠商就會交由公司內部或外包模型師完成樣本製作，但是長谷川的女性人物模型連塗裝都交由辻村聰志本人處理。

「我們認為直到塗裝完成才是辻村先生的作品。新商品公告後，大家對於實際在網站上公開塗裝後的商品給予極大的回響。關於塗裝委託只給予大略的要求說明，細節變化都是由辻村先生處理。若有看到實際成品就可以了解其完美的表現力，我覺得辻村先生在塗裝時留意了顏色的運用，使商品樣本在照片中呈現更漂亮的樣子。」（長谷川）

的確我們消費者最終決定是否會「想要購買」，是根據看到的完成狀態。越了解辻村聰志創造的人物模型魅力，越可以感受到除了人物模型造型之外，直到塗裝修飾的完成狀態才是商品最美的樣子。這代表沒有其他成品會勝過由辻村聰志塗裝完成的樣本。而且長谷川從當初將公司的商品原型交由辻村聰志處理後，就一直未曾改變這種合作模式。這似乎是非常了解辻村聰志創作的魅力而下的判斷。

「若要以廠商的視角來討論辻村造型的魅力，首先成品完成度之高無需贅言，而且工作效率之快非常有利於我們作業。若提到魅力似乎有舉不完的例子，不過辻村先生製作的女性群像總讓人覺得很雅緻。即便是不好

▲2018年1月推出的「開心蛋機少女羽澄麗」，和1/24比例汽車模型＋樹脂女性人物模型系列的展開幾乎是同時期推出的作品。長谷川將人氣題材「蛋機少女」開發成最新系列時，身為初期原型師的辻村聰志左右了日後的商品走向。

This 1/24 scale figure of Rei Hazumi was the pilot product of the soon-to-be Hasegawa's super-popular "Tamago Girls Collection" series.

▲2018年3月銷售的「蛋機少女系列No.01『羽澄麗』w/F-2」。這是附有1/20比例人物模型的人氣系列，辻村聰志負責了系列No.1到No.11的原型製作。這時1/24比例的人物模型系列都已製作到No.5，令人吃驚的是辻村聰志同時負責不同題材系列的原型製作。

After the success of the pilot product, Tsujimura continued to work on No.1 to No.11 entries to the "Tamago Girls Collection" series.

▶2021年10月銷售的1/12比例「女騎士Vol.2」。在系列第1彈的女騎士之後睽違了1年半，以廣告標語「最適合搭配1/12比例摩托車模型的組合」宣布回歸。正因為公司也有推出1/12比例摩托車模型的商品系列，才會選擇這個項目。

The 1/12 scale "Girls Rider Vol.2" is expected to be released in October 2021. These rider figures are ideal for displaying next to 1/12 motorcycle kits.

處理的主題，他都能做出很雅致的品味，讓人毫不猶豫地就做成商品推出。」（長谷川）

即使是性感主題，也不會引來反感，這就是最重要的地方。原因很簡單，就是辻村聰志希望依照自己看起來覺得「美」的樣子化為有形呈現在大眾眼前，因此總是謹慎拿捏構思。人物呈現的模樣並非以男性視角包含慾望和理想等不純粹的念頭，而是單純想要勾勒出線條和輪廓的美感。因此即便女性看到辻村聰志的作品，都會很直覺表示「好可愛」、「好漂亮」。

最後想請問今後長谷川和辻村聰志攜手創作的女性人物模型的發展計劃。

「現在主要商品為廣受好評的性感路線，但是公司內外對於擬真人物模型都有不同的意見需求。或許會參考這些需求做出不同現在路線的商品系列。例如2021年10月銷售的1/24比例『鈴木JIMNY w/露營女性人物模型』，就是因應公司社長喜歡戶外活動的期望而做出的商品。日後或許還會再推出這類不同屬性的品項。我們期待大家提供的想法和意見。」（長谷川）

回顧至今的商品系列，也曾出現過酒店女公關這類讓人吃驚又匪夷所思的品項，主題選擇之出人意表，讓不少消費者都感到驚訝不已。不過或許在不久的將來，女性人物模型會出現更令大眾意想不到的主題。

An example of the product specification sheet given to Tsujimura from Hasegawa at the initial stage of the project. Information such as their hairstyles and facial features are provided in a separate sheet.

◀在初期階段長谷川會給辻村聰志一份商品規格書。除了姿勢，還包括髮型、服裝、臉部長相，而且還會在附件詳細寫上分割提案，由此可知在企劃初始就已經具體列出商品內容細項。辻村聰志可以根據這份資料更具體做出姿勢、表情和成品形象。

Multiple corrections and modifications to the original are made throughout the sculpting process. This process is essential, especially for injection-plastic kits with their mold limitations.

◀除了用照片之外，還會用實物討論原型調整的指示。尤其塑膠製的人物模型指示中還有關於模具的脫模方向等內容，和樹脂製人物模型相比，有更多實際作業的限制。不降低細節，也不破壞形狀，雙方不斷尋求更佳表現持續製作。

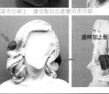

◀2021年10月銷售的1/24比例「鈴木JIMNY w/露營女性人物模型」。相同的人物模型系列也會出現極為不同的主題類型，因此今後的發展真是難以預測，卻也令人期待。

Prototype of a girl figure in a camping outfit expected to be released in October 2021 along with 1/24 scale Suzuki Jimny kit.

◀塗裝指示書。除了色調之外，連材質都預先設定，甚至詳細寫下每種材質呈現的光澤感，由此可知廠商在商品完成形象上有很明確的設定。接著辻村聰志的工作就是消化這些資訊後，在不偏離形象設定的前提下，添加一些原創性並且製作出塗裝樣本。

Here is an example of a painting instruction sheet. Not only the color but also the texture of each material is described in detail. It's Tsujimura's job to digest this information and output it as a painted sample.

ROOTS PART 2

確立「人物模型造形師」
地位的基礎

Following Tsujimura's journey of becoming a master sculptor

撰文／寒河江雅樹　Text：Masaki Sagae

辻村聰志如今因為長谷川女性人物模型原型更廣為人知，然而在開始以商業原型師身分開始活動的前後，又有那些關於造型創作的經歷呢？

辻村聰志在2006年參加了田宮主辦的模型改造大賽，以對自己影響甚鉅的『風之谷娜烏西卡』一舉獲得銅賞。隔年在相同比賽中，以創作歌手坂井泉水為主題的作品再次獲得金賞的殊榮，漸漸以造形師的身分嶄露頭角。之後的一段時間，他慢慢學習累積了市售人物模型改造等造型相關知識與經驗。2007年起一邊製作博物館展示用的模型，一邊製作原創人物模型和偶像應援的人物模型。另外，他個人還以模型師的身分參與立體創作的活動，或是在網路拍賣銷售造型作品，雖然只是以個人名義持續參與造型活動，但是漸漸在人物模型領域打響了知名度。因此經過朋友的介紹，從2014年開始製作商業原型，也就是MODELKASTEN 1/35比例女高中生的人物模型系列。這個系列不同於現在長谷川的原型委託，除了女高中生主題之外，幾乎沒有其他相關指示，姿勢和細節等都由辻村聰志自由發揮。這一個系列引起大眾的討論後，深受市場的熱烈歡迎，甚至有部分品項在之後做成射出成型的塑膠模型發售（現在已停止販售）。而且辻村聰志透過這個系列的製作，在原型製作方面累積不少的技巧。至今製作偶像人物模型的造型實力獲得大眾的認可，在2015年受到「人物宛如本尊」的好評，而和海洋堂合作了『水中過膝長襪美少女真縞花栗鼠』的原型製作。辻村聰志在和長谷川合作前就因為商業原型製作廣為人知，例如：配合攝影創作者古賀學『水中過膝長襪美少女立方體展（2015年8月舉辦）』於現場販售人物模型，還在網路新聞受到極大的關注。接著如同前面所說在2014～2017年在多家廠商的商業原型製作和立體創作活動中都可看到他的身影，而自2017年至今則大多從事長谷川商品的原型製作。

Tsujimura won his first award in the 35th Figure Modification Contest. His figure, which depicted the main character from the movie "Nausicaa of the Valley of the Wind" has won the bronze prize. He then proceeded to win the gold award the following year. Tsujimura's popularity boosted after working on a figure of a well-known Japanese actress Rina Akiyama. His work was spread all around Japanese media via social networking services, and the completed figure was eventually gifted to Akiyama herself. Tsujimura's love for idols is well known, and he has often made figures of them as well. Just like his paintings, Tsujimura sculpts figures to make people joyful, to make people appreciate the beauty of a person. Even as a commercial and professional sculptor, his dedication to making people happier has not changed. Whenever he is working on a figure, Tsujimura hopes to input as much beauty as he can. He sincerely wishes everybody to like his works and to add their personal touch to his creations.

5

7

8

13

9

10

11

12

6

1 水中過膝長襪美少女真縞花栗鼠樹脂套件（1/10比例，於2015年製作）。海洋堂和辻村聰志合作將水中寫真集『水中過膝長襪美少女』系列中穿著過膝長襪的美少女做成人物模型。樹脂套件除了角色扮演者真縞花栗鼠的表情，連潛水鏡、蛙鞋和呼吸調節器等裝備細節都完整重現。

2 附臀部的半身像（1/10比例，大概於2005年製作）。半身像通常是指胸部以上的部分，但是想試看看保留美麗的臀部線條，而且似乎也沒有人做過，辻村聰志基於這個想法而練習試作。

3 這個1/35比例的娜烏西卡是2006年在第35屆模型人偶改造大賽中獲得銅賞的作品。辻村聰志之前就一直改造1/35比例的人偶，但卻是第一次參加這個比賽。當時還有「隆美爾和幕僚軍官組」等共 3 件作品參加比賽。

4 美臀秋山莉奈，也就是女星秋山莉奈（於2014年製作）。某種程度上這是辻村聰志的成名之作，讓他在擬真人物模型造型的實力廣為人知。而且當時正值SNS等作品上傳風潮興起並迅速擴散。這件作品還贈送給秋山莉奈本人。

5 Maria（1/20比例，於2005年製作）。製作靈感來自穿著這套設計泳裝的國外模特兒。作品想表現的是臀部的優美線條，也有在立體創作活動中販售。

6 身著乳膠裝的女郎（1/24比例，大概於2006年製作）。這件原創作品想表現設計的樂趣和女性的柔美線條，並且也有在立體創作活動中限定販售。

7 Black Daddy（1/24比例，於2006年製作）。作品改造自ANDREA公司白色金屬套件的臉和頭髮。

8 背著書包的女孩（1/35比例，大概於2007年製作）。並沒有依據的模特兒，一開始製作時就想盡量做小一些，在造型作業上雖然有趣，但是塗裝時卻很費工。

以下刊登的人物模型是由辻村聰志製作原型的MODELKASTEN商品，部分在2021年依舊可以購得。

MODELKASTEN線上商店
（http://store.modelkasten.com/shopbrand/figure/）

9 JK-01/Sayaka學姊（高中3年級）（1/35比例，含稅2750日圓，樹脂鑄造套件）2014年製作，MODELKASTEN女高中生人物模型系列第1彈。呈現動態姿勢，女高中生身穿夏季制服，稍微提起沉重的裙子邁開步伐。

10 JK-09/ Yuuka（高中3年級）（1/35比例，含稅2970日圓，樹脂鑄造套件）MODELKASTEN女高中生人物模型系列第 9 彈。女高中生的主題設定為，居住在蘆屋市高級住宅區，身穿包裹全身的冬季制服，只露出一張臉。有點重的學生書包為細節亮點。

11 JK-12/ Sumire（高中2年級）（1/35比例，含稅2860日圓，樹脂鑄造套件）MODELKASTEN女高中生人物模型系列第12彈。作品形象為表情有點成熟的少女，坐在夕陽照入的和室中。造型呈現許多亮點，包括坐姿、衣服皺褶和重疊的雙手。

12 JK-13/ Rumi（高中3年級）（1/35比例，含稅3080日圓，樹脂鑄造套件）MODELKASTEN女高中生人物模型系列第13彈。最後一天身穿水手服，揮別開心的高中生活，造型重點為落寞的神情和手拿的畢業證書筒。

13 解剖系列No.02（1/35比例，含稅2860日圓，樹脂鑄造套件）人體構造系列的第 2 彈，作品做成身穿泳衣的西方女性，也可以單純活用作素體。馬尾髮型和蝴蝶結為背影添加迷人色彩。

1. 1/10 scale resin figure of Japanese cosplayer Shimarisu Mashima. Even her diving equipment is accurately represented.

2. Typically, a bust figure is cut around its waist, but this sculpture created back in 2005 includes its hips as a bonus.

3. 1/35 scale figure of Nausicaa, which won a bronze prize in the 35th Figure Modification Contest (2016) hosted by Tamiya.

4. A figure of famous Japanese actress Natsuki Akiyama. Made in 2014, this figure boosted Tsujimura's popularity via SNS.

5. After being inspired by a model wearing the same bikini, Tsujimura created this 1/20 scale figure in 2005.

6. A girl in a rubber suit, sculpted in 1/24 scale in 2006. Some of these figures were cast and sold in various conventions.

7. A 1/24 scale figure titled "Black Daddy" created in 2006. A white metal figure from Andrea Miniatures was used as a base.

8. A girl carrying a backpack, 1/35 scale, 2007. Sculpting something as tiny as this was fun, but painting it was a nightmare.

9. This high school girl figure was the first iteration of Model Kasten's "High School Girls Figures Series." 1/35 scale, 2014.

10. The ninth introduction to the same series. You can barely see her face, yet her facial features are easy to read.

11. The twelfth introduction to the same series. Small details such as the creases of her clothing and her hands are out of the world.

12. Another figure from the same series. This girl wears a traditional Japanese-style uniform carrying a graduation certificate.

13. The second entry of the "Anatomy Series," 1/35 scale, 2014. This figure is best used as a base for creating your original figure.

SCULPTURE BEAUTY'S
人物模型之美
辻村聰志　女性人物模型作品集

翻　　譯	黃姿頤
發　　行	陳偉祥
出　　版	北星圖書事業股份有限公司
地　　址	234新北市永和區中正路462號B1
電　　話	886-2-29229000
傳　　真	886-2-29229041
網　　址	www.nsbooks.com.tw
E-MAIL	nsbook@nsbooks.com.tw
劃撥帳戶	北星文化事業有限公司
劃撥帳號	50042987
出 版 日	2022年8月
I S B N	978-626-7062-32-6
定　　價	480元